Contents

UFO Investigation

By John Michael Ashley

UFO Network 2023

Chapter 1 Introduction

Welcome to this guide on how to investigate unidentified flying objects, or UFOs. For decades, people have been fascinated by reports of strange lights in the sky, mysterious craft hovering in the distance, and unexplainable encounters with otherworldly beings. And as UFO sightings continue to be reported around the world, there is a growing need for trained investigators who can collect and analyze the data in a systematic and scientific way.

In this book, we will provide a comprehensive guide to conducting UFO investigations, covering everything from the basics of understanding what UFOs are to the latest scientific approaches to analyzing the data. Whether you are a seasoned investigator or a curious beginner, we hope to provide you with the tools, resources, and knowledge you need to explore this fascinating and complex subject.

Of course, investigating UFOs is not without its challenges. From hoaxes and misidentifications to public skepticism and government secrecy, there are many obstacles that investigators must overcome. But by approaching this subject with an open mind, a critical eye, and a commitment to scientific rigor, we believe that it is possible to uncover important information about the nature of these elusive phenomena.

Throughout this book, we will provide practical advice and guidance on conducting UFO investigations, drawing on the latest research, techniques, and technologies. We will cover topics such as how to collect and analyze UFO data, how to debunk hoaxes and misidentifications, and how to communicate your findings to the public and other interested parties.

But our approach will not be purely technical. We also recognize the importance of ethics, communication, and collaboration in conducting successful UFO investigations. We will explore the challenges and responsibilities involved in investigating potentially sensitive or controversial information, and we will provide guidance on how to approach these issues with sensitivity and professionalism.

Of course, we do not pretend to have all the answers, and there is still much that we do not know about the nature of UFOs. But by sharing our knowledge and experience, we hope to inspire a new generation of investigators and to advance our understanding of these fascinating phenomena.

So if you are ready to embark on a journey of discovery, to explore the mystery and wonder of UFOs, and to contribute to the scientific study of these elusive phenomena, then join us as we dive into the exciting world of UFO investigations.

Chapter 2 Understanding UFOs

Unidentified Flying Objects, or UFOs, are a fascinating and often elusive phenomenon that have captured the imagination of the public and the interest of researchers for decades. The term "UFO" refers to any object or phenomenon that appears to be flying in the sky and cannot be easily identified. While many UFO sightings can be explained as natural phenomena, man-made objects, or hoaxes, a significant number of reports remain unexplained.

One of the biggest challenges in studying UFOs is defining what they are. The term "UFO" has become associated with extraterrestrial visitation or advanced military technology, but it is important to remember that the term itself simply refers to an unidentified flying object. This means that a UFO can be anything that appears to be flying and cannot be easily identified, including natural phenomena like meteors or weather balloons.

To help differentiate between UFOs and identified

aerial phenomena (IAPs), the term "unidentified aerial phenomena" (UAP) has been used by some researchers. This term is used to describe any aerial phenomenon that cannot be readily identified or explained. It is important to note that while some UAPs may be UFOs, not all UFOs are necessarily UAPs.

Eyewitness testimony is one of the most important sources of data for UFO investigations. Witnesses may describe seeing objects in the sky that move in ways that defy the laws of physics, such as sudden changes in direction or speed. However, eyewitness testimony is not always reliable, and it is important for investigators to corroborate witness accounts with other types of data.

Photographic and video evidence can also be valuable sources of data for UFO investigations. Advances in technology have made it easier for people to capture images and videos of objects in the sky, but this evidence is not always conclusive. Photos and videos can be faked or misinterpreted, and it is important for investigators to carefully analyze this type of evidence.

Radar data is another important tool for analyzing UFO sightings. Radar can detect objects in the sky that may not be visible to the naked eye, and it can provide valuable information about an object's speed, altitude, and trajectory. However, radar data can also be affected by a variety of factors, including weather conditions and interference from other objects.

Physical trace evidence, such as soil and plant samples, can also be valuable sources of data for UFO investigations. In some cases, witnesses may report seeing objects land or leave physical traces on the ground. Collecting and analyzing physical evidence can provide clues about the nature of the object and its possible origin.

One of the biggest challenges in investigating UFOs is debunking hoaxes and misidentifications. Some UFO sightings can be explained as natural phenomena, man-made objects like planes or drones, or intentional hoaxes. It is important for investigators to carefully analyze all available data to determine whether a sighting is genuine or a hoax.

Critical thinking and skepticism are important skills for UFO investigators. While it is important to remain open-minded and impartial, it is also important to approach UFO sightings with a healthy dose of skepticism. This means carefully analyzing all available data and considering alternative explanations before jumping to conclusions.

The history of UFO sightings and their impact on society is also an important topic for UFO investigations. UFO sightings have been reported for centuries, and they have often been associated with religious or spiritual beliefs. The way that

UFO sightings are perceived by the public and the media can have a significant impact on how they are studied and understood.

The psychological and sociological aspects of UFO belief are also important topics for investigation. Why do some people believe in UFOs and others do not? How does belief in UFOs impact people's beliefs and attitudes about science, religion, and society? Understanding these factors can help investigators with their work.

Another important concept in understanding UFOs is the phenomenon of "close encounters." The late astronomer and UFO researcher J. Allen Hynek developed a classification system for these encounters, ranging from sightings of UFOs at a distance to actual physical contact with extraterrestrial beings. This system is widely used in the field of ufology and provides a framework for categorizing and analyzing UFO reports.

It's also essential to consider the psychological and social factors that influence UFO sightings and reports. People's expectations, beliefs, and prior experiences can all affect how they perceive and interpret unusual events in the sky. Additionally, cultural and societal factors, such as media coverage and the prevalence of science fiction, can shape people's attitudes toward UFOs and influence their willingness to report sightings.

In recent years, there has been a growing interest

in the scientific study of UFOs, often referred to as "UAP" (unidentified aerial phenomena) by the U.S. government and military. This approach involves gathering data using advanced technologies such as radar and infrared cameras, as well as analyzing witness accounts and physical evidence. The goal is to identify patterns and characteristics of UAPs that could provide clues about their nature and origin.

Another aspect of understanding UFOs is exploring their potential connection to extraterrestrial life. While the existence of alien life has not been confirmed, the vastness of the universe and the discovery of habitable planets make the possibility intriguing. Some researchers believe that UFO sightings could be evidence of visits by advanced extraterrestrial civilizations, while others propose that these sightings are part of a larger phenomenon that has yet to be fully understood.

Ultimately, understanding UFOs requires an open-minded but critical approach that seeks to gather and analyze data while remaining aware of potential biases and preconceptions. By exploring the history, science, and societal implications of UFOs, investigators can develop a comprehensive understanding of this fascinating and mysterious phenomenon.

Chapter 3 Collecting Data

Collecting UFO data is a crucial part of any UFO investigation. To effectively investigate a UFO sighting, it's important to collect as much information as possible from witnesses and physical evidence. In this section, we'll explore various methods for collecting UFO data and the best practices for doing so.

One of the most valuable types of data in UFO investigations is detailed witness testimony. Encouraging witnesses to provide as much detail as possible, including the date, time, location, and duration of the sighting, the size and shape of the object, and any unusual movements or sounds, is crucial for a thorough investigation.

In addition to eyewitness accounts, collecting physical evidence can also provide valuable insights into the nature of UFOs. Landing marks, crop circles, and radiation traces are just a few examples of physical evidence that investigators may encounter during an investigation. Properly collecting and analyzing this evidence can help investigators identify the possible origin of the sighting.

Using advanced technologies such as radar, infrared cameras, and other sensors can also provide valuable data in a UFO investigation. These technologies can help identify and track UFOs, and provide additional information about the object's speed, altitude, and other characteristics.

Interviewing witnesses is another important method for collecting UFO data. Conducting interviews in a non-judgmental and objective manner is crucial to avoid bias or leading the witness. It's important to ask open-ended questions and avoid suggesting explanations for the sighting.

Establishing a database of UFO sightings is also an essential part of collecting UFO data. This database can help identify patterns or trends over time and provide valuable information for analysis and research.

Collaborating and sharing data with other UFO investigators is beneficial in corroborating sightings and analyzing data. By working together, investigators can identify patterns and trends that may not be apparent from individual investigations.

Maintaining confidentiality is crucial when collecting UFO data. Witnesses may be hesitant to come forward with their sighting due to fear of ridicule or public attention. It's important to respect their confidentiality and provide a safe space for them to report their experience.

It's also important to be mindful of ethical considerations when collecting UFO data. Informed consent is crucial when interviewing witnesses, and investigators should avoid manipulating witness testimony to fit a particular narrative.

In addition to eyewitness testimony, physical evidence, advanced technologies, interviewing witnesses, establishing a database, collaborating with other investigators, maintaining confidentiality, and ethical considerations, investigators should also be prepared to handle the emotional impact of a UFO sighting. Witnesses may be shaken or traumatized by their experience, and it's important to approach the investigation with empathy and understanding.

Overall, collecting UFO data is a crucial part of any investigation. By using a combination of methods and best practices, investigators can collect the most accurate and reliable data possible, leading to a deeper understanding of the UFO phenomenon.

Chapter 4 Analyzing Data

Data analysis is a crucial step in any UFO investigation. In this section, we will explore the methods, techniques, and challenges involved in analyzing the data collected during UFO investigations.

The first step in analyzing UFO data is to carefully evaluate each piece of evidence to determine its reliability and relevance to the investigation. This may involve interviewing witnesses to gather additional information or examining photographs and videos to determine their authenticity.

Once the data has been collected and evaluated, analysts can use a variety of methods and techniques to analyze it. One common approach is statistical analysis, which involves using mathematical models to identify patterns or anomalies in the data.

Another approach is pattern recognition, which involves identifying similarities and differences between different UFO cases to identify common characteristics or trends. This can help investigators develop hypotheses or theories about the nature of

UFO sightings and encounters.

Triangulation is another useful technique for analyzing UFO data. This involves using multiple sources of data, such as eyewitness testimony, radar data, and physical evidence, to corroborate or verify the accuracy of each individual data point.

Technology can also play a key role in analyzing UFO data. Computer software, such as mapping tools or image analysis software, can help investigators visualize and analyze the data more effectively.

Experts and specialists in specific fields, such as photography or radar analysis, may also be consulted to help analyze specific types of data. Their expertise can help ensure that the data is interpreted correctly and accurately.

There are also several challenges and limitations associated with analyzing UFO data. For example, witness testimony may be subjective and open to interpretation, and physical evidence may be contaminated or otherwise compromised.

Additionally, it can be difficult to distinguish genuine UFO sightings from hoaxes or misinterpretations, and technological limitations may prevent investigators from collecting certain types of data or analyzing them effectively.

Despite these challenges, it is important to document and archive UFO data for future analysis and comparison with other cases. By maintaining a database of UFO data, investigators can identify patterns and trends over time and gain a better understanding of the nature of UFO sightings and encounters.

Without a doubt, analyzing UFO data is a complex and challenging process that requires careful evaluation, rigorous analysis, and the use of advanced technology and expertise. By following best practices and utilizing a variety of techniques and methods, investigators can better understand the nature of UFO sightings and encounters and make informed conclusions about their origin and meaning.

Chapter 5 Debunking Hoaxes

Despite the abundance of legitimate UFO sightings, there have been countless hoaxes throughout history. The motives behind these hoaxes vary from individual to individual, but they can have serious consequences for the credibility of UFO research. In this section, we will explore the topic of UFO hoaxes and how to differentiate them from genuine sightings.

One common characteristic of UFO hoaxes is the use of blurry or poorly-lit photographs or videos. In many cases, these images are intentionally created to be difficult to discern, making it hard to verify their authenticity. UFO hoaxes may also rely on sensational or outlandish claims, often accompanied by a lack of verifiable evidence.

It is important to properly investigate and analyze UFO sightings in order to distinguish hoaxes from genuine sightings. One key aspect of this investigation is to determine whether there is any corroborating evidence, such as multiple witness accounts or physical evidence at the site of the sighting. It is also important to carefully scrutinize the background and history of the person making the claim, as this can reveal any potential biases or motives for perpetuating a hoax.

UFO hoaxes can have serious consequences, including the spread of misinformation and the discrediting of legitimate UFO research. The media can play a role in perpetuating and debunking UFO hoaxes, and it is important for journalists and news outlets to carefully vet their sources and the information they present.

Critical thinking and skepticism are important tools for evaluating UFO claims and distinguishing hoaxes from genuine sightings. In some cases, hoaxes can be perpetuated for financial gain or attention-seeking purposes, while in other cases they may be part of a larger disinformation campaign.

Legal and ethical implications can also arise from perpetrating UFO hoaxes. Depending on the nature of the hoax, individuals may be subject to criminal charges or civil lawsuits for the harm caused by their actions. In addition, perpetuating a hoax can have serious consequences for the reputation of the individual or organization involved.

Investigators and researchers have a responsibility to debunk hoaxes and promote factual information. This includes engaging with the media and the public to help distinguish between genuine sightings and hoaxes, as well as conducting rigorous investigations and analysis to properly identify hoaxes.

While the motivations behind UFO hoaxes can vary widely, it is important to remain vigilant and skeptical when evaluating UFO claims. By carefully scrutinizing the evidence and remaining objective in our analysis, we can help separate fact from fiction and better understand the genuine sightings that may hold important clues to the existence of extraterrestrial life.

Chapter 6 Historical Cases

Throughout history, there have been numerous sightings and encounters with unidentified flying objects (UFOs) that have captured the public's attention and imagination. In this section, we will explore some of the most significant and intriguing historical UFO cases.

One of the most famous cases occurred in June 1947, when private pilot Kenneth Arnold reported seeing a group of nine disc-shaped objects flying in formation near Mount Rainier in Washington State. This sighting, which garnered widespread media attention, is credited with popularizing the term "flying saucer."

Another highly publicized case was the 1947 Roswell incident, in which a rancher in New Mexico claimed to have found debris from a crashed UFO on his property. While the government initially stated that the debris was from a weather balloon, many people remain convinced that it was evidence of extraterrestrial life and a subsequent cover-up.

In September 1952, a series of UFO sightings occurred over Washington D.C., which resulted in the Air Force conducting an investigation. While the government ultimately claimed that the sightings were due to a temperature inversion, some researchers and witnesses maintain that the objects were extraterrestrial in nature.

One of the most well-known abduction cases involves Betty and Barney Hill, who claimed to have been taken aboard a UFO in 1961 while driving through rural New Hampshire. The case was heavily investigated and remains one of the most studied and debated abduction cases to this day.

In 1975, logger Travis Walton was allegedly abducted by a UFO while working in Arizona's Apache-Sitgreaves National Forest. His story gained widespread media attention and was later adapted into a Hollywood movie.

Another intriguing case occurred in March 1997, when thousands of people in Phoenix, Arizona reported seeing a massive V-shaped UFO in the sky. While the government claimed that the object was simply flares from military exercises, many witnesses maintain that it was a genuine UFO.

The Rendlesham Forest incident, which occurred in December 1980 near a U.S. Air Force base in England, is often referred to as Britain's Roswell. The incident involved multiple witnesses seeing strange lights and a triangular-shaped craft in the woods near the base.

In 1989 and 1990, the small European country of Belgium experienced a wave of UFO sightings that involved triangular-shaped craft. The sightings were widely reported in the media and investigated by the Belgian Air Force.

South America has also had its share of UFO activity, with one of the most notable cases occurring in Chile in 1977. The government launched an official investigation into the sighting, which involved multiple witnesses seeing a disc-shaped object over the city of Valparaíso.

These are just a few examples of the historical UFO cases that have captured the public's attention over the years. While many of these cases have been investigated and explained away, others remain a mystery and continue to fuel speculation and debate.

Chapter 7 Current Cases

As technology advances and more people have access to high-quality recording devices, reports of UFO sightings are becoming increasingly common. In this section, we'll explore some of the most notable recent UFO cases, including the ongoing efforts of the Galileo Project to investigate these phenomena.

One recent case that gained widespread attention was the 2019 Navy UFO sightings, in which several pilots reported seeing unidentified flying objects that appeared to defy the laws of physics. Despite the Navy initially stating that the objects seen in the footage were "unidentified aerial phenomena," some skeptics have suggested that they could be explainable by natural phenomena or advanced military technology.

Another intriguing case is the Phoenix Lights incident of 1997, in which thousands of people reported seeing a massive triangular object flying over the city. Although the military initially denied any involvement, some have suggested that the object was a secret government aircraft or even an extraterrestrial craft.

In recent years, the Galileo Project has emerged as a leader in the field of UFO investigation. Led by Harvard astrophysicist Avi Loeb, the project aims to use cutting-edge technology to search for evidence of extraterrestrial intelligence. The team plans to use telescopes, cameras, and other equipment to monitor the skies for signs of unusual activity.

One key area of focus for the Galileo Project is the search for interstellar objects like 'Oumuamua, the mysterious cigar-shaped object that passed through our solar system in 2017. Loeb has suggested that 'Oumuamua could have been an alien spacecraft, prompting widespread debate and speculation within the scientific community.

Other recent UFO cases include the TikTok UFO footage, which appears to show a spherical object moving at incredible speeds, and the 2018 Aguadilla Airport incident, in which a mysterious object was spotted hovering over the airport before disappearing into the ocean.

Despite the growing body of evidence for the existence of UFOs, many skeptics remain unconvinced. Some argue that most UFO sightings can be easily explained by natural phenomena or human error, while others suggest that the phenomena may be the result of psychological or cultural factors.

Regardless of one's personal beliefs about the nature of UFOs, it's clear that the topic will continue to be a subject of fascination and inquiry for years to come. With new technologies and research methods emerging all the time, there's no telling what exciting new developments we may uncover in the field of UFO investigation in the years ahead.

Chapter 8 UFO Abduction

UFO abductions are a controversial and fascinating aspect of the UFO phenomenon. While some dismiss abduction claims as mere hoaxes or delusions, others consider them to be some of the most compelling evidence of alien visitation.

Abduction reports typically involve a person claiming to have been taken aboard a UFO against their will and subjected to various procedures and examinations. Some claim to have been physically and sexually assaulted, while others report more benign interactions.

One of the most famous abduction cases is that of Betty and Barney Hill, who claimed to have been taken aboard a UFO while driving through rural New Hampshire in 1961. Their detailed descriptions of their experience, which included medical examinations and communication with the beings, sparked widespread interest in UFO abductions.

Another well-known case is that of Travis Walton, who claimed to have been taken aboard a UFO while working in the Apache-Sitgreaves National

Forest in Arizona in 1975. Walton's disappearance and subsequent reappearance five days later, along with his vivid descriptions of his experience, have made his case one of the most compelling abduction stories.

Many abduction researchers believe that these experiences are not just isolated incidents, but rather part of a larger phenomenon. They point to similarities in the stories told by abductees, such as the presence of beings with large heads and eyes, and the use of medical procedures.

Some skeptics argue that abduction experiences are simply the result of sleep paralysis or other natural phenomena. However, others argue that the consistency of the stories across cultures and time periods suggests that there may be something more to these claims.

Despite the controversy surrounding abduction claims, many researchers continue to investigate these reports in an effort to better understand the UFO phenomenon. The Galileo Project, for example, plans to investigate potential extraterrestrial technological artifacts, including those that could potentially have been used in abductions.

While there is still much to learn about UFO abductions, it is clear that these reports have captured the imagination of many and continue to be a subject of intense interest and investigation. Whether they are the result of psychological phenomena, extraterrestrial visitation, or something else entirely, the mystery of UFO abductions remains an intriguing and compelling topic of study.

Chapter 9 Government and Military Involvement

Since the beginning of modern UFO sightings, many have speculated about the government and military's involvement in investigating these sightings. Conspiracy theories abound, with some believing that the government is hiding information about UFOs from the public. But what is the truth behind these speculations, and how can UFO investigators navigate this complex and potentially fraught area?

To start, it's important to understand the historical context of government and military involvement in UFO investigation. Beginning in the 1940s and continuing through the Cold War, the U.S. Air Force conducted a series of investigations into UFO sightings, including Project Sign, Project Grudge, and Project Blue Book. While the results of these investigations were often inconclusive, they demonstrate the government's interest in understanding these phenomena.

Today, the government's involvement in UFO

investigation is less overt, but still exists in various forms. For example, in June 2021, the U.S. government released a report detailing numerous unexplained sightings by military pilots. The report indicates that the government takes these sightings seriously and is actively investigating them.

Given this context, UFO investigators should be prepared to encounter government and military personnel in the course of their work. It's important to approach these encounters with professionalism and respect, while also recognizing that there may be limitations on the information that government officials are able or willing to share.

UFO investigators should also be aware of the legal implications of investigating sightings on government or military property. Trespassing on military installations is illegal and can result in serious consequences, including arrest and prosecution.

In addition, investigators should be familiar with the Freedom of Information Act (FOIA), which allows individuals to request access to government documents. While the government may withhold certain information for national security reasons, FOIA requests can still be a valuable tool for investigators seeking to obtain information about UFO sightings.

Another important consideration for UFO investigators is the role of intelligence agencies in

UFO investigation. The Central Intelligence Agency (CIA) has acknowledged its interest in UFOs and has released previously classified documents related to UFO sightings. However, it's important to remember that intelligence agencies may have their own agendas and may not necessarily be forthcoming with information.

Finally, it's worth noting that some UFO sightings may be the result of secret military technology rather than extraterrestrial visitors. Investigators should be prepared to consider this possibility and to look for evidence of military involvement in sightings.

In summary, government and military involvement in UFO investigation is a complex and multifaceted topic. While there may be limitations on the information that investigators are able to obtain, it's important to approach encounters with government and military personnel with professionalism and respect. By understanding the historical context of government involvement in UFO investigation, as well as the legal and ethical considerations involved, investigators can navigate this area of investigation with confidence and integrity.

Chapter 10 The Scientific Approach

The scientific approach is a methodical process that involves observation, hypothesis formulation, experimentation, and analysis of data. This method can be applied to the investigation of UFO sightings, with the goal of collecting empirical evidence that can help us understand the nature of these phenomena. In this chapter, we will explore various scientific approaches to UFO investigation, including data collection, analysis, and experimentation.

1. Data Collection
The first step in any scientific investigation is the collection of data. In the case of UFO sightings, this can involve eyewitness accounts, photographs, videos, radar data, and other forms of physical evidence. The challenge is to collect accurate and reliable data that can be analyzed scientifically. To achieve this, investigators need to follow strict protocols for data collection and ensure that their methods are transparent and replicable.

2. Analysis of Data

Once data is collected, it needs to be analyzed to draw meaningful conclusions. This involves statistical analysis, triangulation of different sources of data, and consideration of alternative hypotheses. The goal is to identify patterns or anomalies in the data that can shed light on the nature of the phenomenon under investigation.

3. Experimentation

Experimentation is a crucial aspect of the scientific method, and it can be applied to UFO investigation. For example, investigators can conduct experiments to test hypotheses about the physical properties of UFOs, such as their speed, size, or luminosity. These experiments can involve the use of radar, telescopes, or other scientific instruments to gather data under controlled conditions.

4. Hypothesis Formulation

One of the main goals of scientific investigation is to develop testable hypotheses that can explain observed phenomena. In the case of UFOs, there are numerous hypotheses that have been proposed, ranging from extraterrestrial visitation to advanced military technology to natural phenomena. The challenge is to develop hypotheses that are testable, falsifiable, and grounded in scientific evidence.

5. Peer Review

Peer review is an essential aspect of the scientific method, whereby scientific findings are evaluated by independent experts in the field. This process helps to ensure that scientific research is of high quality and free from bias. In the case of UFO investigation, peer review can help to validate or refute claims made by investigators.

6. Replicability
Replicability is a crucial aspect of scientific investigation, whereby experiments and analyses are repeated by independent researchers to confirm the validity of findings. In the case of UFO investigation, replicability is challenging due to the rarity and unpredictability of UFO sightings. However, efforts can be made to develop standardized protocols for data collection and analysis to increase the replicability of investigations.

7. Integration with Other Disciplines
UFO investigation can benefit from integration with other disciplines, such as physics, astronomy, and psychology. For example, physicists can help to analyze the physical properties of UFO sightings, astronomers can provide insights into the astronomical context of sightings, and psychologists can investigate the perceptual and cognitive factors that influence eyewitness testimony.

8. Skepticism
Skepticism is a key element of the scientific method, whereby claims are subjected to critical evaluation

and scrutiny. In the case of UFO investigation, skepticism is essential to ensure that claims are based on scientific evidence and not on wishful thinking, misinterpretation, or hoaxes. Skeptical inquiry can help to identify flaws in arguments and evidence and lead to a more accurate understanding of the phenomenon.

9. Open-Mindedness
While skepticism is essential, investigators must also be open-minded and willing to consider alternative explanations for UFO sightings. The scientific method requires that hypotheses be tested against alternative hypotheses, and that evidence be evaluated objectively. By being open-minded, investigators can avoid the pitfalls of bias and dogmatism, and arrive at a more accurate understanding of the phenomenon.

In addition to analyzing physical evidence, scientists can also use mathematical models to help explain UFO sightings. For example, some sightings have been explained by atmospheric phenomena such as ball lightning or plasma, which can be modeled using mathematical equations. Other sightings have been explained as misidentifications of natural phenomena, such as stars, planets, or aircraft.

However, it is important to note that not all UFO sightings can be easily explained through scientific analysis. Some sightings may remain unexplained due to lack of data, unreliable witnesses, or the limitations of current scientific knowledge. In such cases, scientists may need to consider alternative hypotheses or theories, such as the extraterrestrial hypothesis or the interdimensional hypothesis.

Overall, the scientific method provides a rigorous and systematic approach to investigating UFO sightings. By collecting and analyzing data, using objective criteria, and testing hypotheses, scientists can gain a deeper understanding of these phenomena and contribute to the advancement of scientific knowledge. As our understanding of the universe and technology continues to evolve, it is important to approach UFO investigations with an open mind and a willingness to consider multiple perspectives and explanations.

Chapter 11 Spiritual and Metaphysical Perspectives on UFOs

UFOs have fascinated and mystified people for decades, and for some, their existence has significant spiritual and metaphysical implications. While some may dismiss such ideas as mere fantasy, there are those who believe that the UFO phenomenon represents a powerful spiritual force that has the ability to transform human consciousness. In this chapter, we will explore some of the spiritual and metaphysical perspectives on UFOs, and how they can shape our understanding of this fascinating phenomenon.

One of the most common spiritual interpretations of UFOs is that they are signs or messages from a higher power or extraterrestrial beings. Many people have reported experiencing profound spiritual or mystical experiences during UFO encounters, which have led them to believe that the beings behind the sightings are benevolent and possess a higher spiritual knowledge. Some even suggest that these encounters represent a kind of spiritual awakening, where the experiencer is given a glimpse into a larger, more meaningful reality.

Another perspective is that UFOs represent a kind of spiritual test, where humanity is being challenged to evolve to a higher level of consciousness. Some UFO researchers argue that the beings behind the sightings are trying to awaken us to the interconnectedness of all things, and to the need for greater love, compassion, and understanding in the world.

From a metaphysical standpoint, some believe that UFOs are manifestations of other dimensions or parallel realities. They may represent a bridge between the physical and spiritual worlds, and offer us a glimpse into the true nature of reality. Some UFO researchers suggest that these sightings could be evidence of interdimensional travel, where beings from other realms or parallel dimensions are visiting our world.

Others believe that UFOs have a more ominous side, and may represent a darker, malevolent force. Some suggest that they could represent a kind of spiritual deception, where the beings behind the sightings are manipulating us for their own purposes. This perspective suggests that we need to be cautious in our interactions with these beings, and that we should not assume that they are always benevolent or have our best interests at heart.

Despite the many interpretations and perspectives on UFOs, one thing is certain: they continue to fascinate and intrigue people from all walks of life. Whether they represent a spiritual awakening, a test of humanity's evolution, or something more sinister, the UFO phenomenon has captured our imaginations and sparked our curiosity.

As we continue to investigate this fascinating phenomenon, we must remain open to all possibilities and perspectives, and strive to keep an open mind. By doing so, we may uncover new insights and truths about the nature of our reality, and the role that UFOs may play in it.

Chapter 12 Contact with Extraterrestrial Life

In the realm of UFO investigation, the idea of contact with extraterrestrial life is both exhilarating and daunting. The possibility of making contact with intelligent beings from beyond our planet has captivated human imagination for centuries. But as investigators, it is important to approach this topic with an open mind, tempered by critical thinking and scientific methodology.

One of the challenges of investigating claims of contact with extraterrestrial life is the lack of tangible evidence. Many reports rely solely on personal accounts and anecdotal evidence, which can be difficult to verify. However, this does not mean that the investigation of such claims is futile. By using scientific approaches, we can gather and analyze the available evidence, and draw conclusions based on sound reasoning.

A key element in investigating claims of contact with extraterrestrial life is discernment. Not all reports of contact are genuine, and many can be explained by natural phenomena or psychological factors. As investigators, it is our responsibility to separate fact from fiction, and to approach each case with a healthy dose of skepticism. This doesn't mean that we should dismiss every report outright, but rather that we should approach each case with an objective and impartial mindset.

One approach to investigating claims of contact with extraterrestrial life is to look for corroborating evidence. This might include physical traces left behind by the alleged spacecraft or beings, or witness testimony from multiple sources. The more evidence we can gather, the more confident we can be in our conclusions.

Another important aspect of investigating claims of contact is to consider the motivations of those making the claims. While some may genuinely believe that they have had contact with extraterrestrial life, others may have ulterior motives such as fame, financial gain, or personal vendettas. It is important to approach each case with a healthy skepticism, and to investigate the motivations of all parties involved.

Perhaps one of the most exciting aspects of investigating claims of contact with extraterrestrial life is the potential for new discoveries and insights. As we continue to explore the cosmos and expand our knowledge of the universe, the possibility of making contact with intelligent beings becomes increasingly plausible. By approaching this topic with a scientific mindset, we can ensure that any claims of contact are thoroughly investigated and evaluated.

One of the challenges of investigating claims of contact is the societal stigma surrounding the topic. Many people are hesitant to come forward with their experiences for fear of ridicule or social ostracism. As investigators, it is important to create a safe and supportive environment for those who wish to share their stories. This can be achieved by approaching each case with empathy and understanding, and by respecting the privacy and confidentiality of witnesses.

Another important consideration when investigating claims of contact with extraterrestrial life is the potential impact on society. The discovery of intelligent extraterrestrial life would be a paradigm-shifting event, with far-reaching implications for our understanding of ourselves and our place in the universe. As investigators, it is important to approach this topic with a sense of responsibility, and to consider the potential consequences of any findings.

In conclusion, investigating claims of contact with extraterrestrial life is a complex and challenging endeavor, but one that is essential to the advancement of our understanding of the universe. By approaching this topic with a scientific mindset, tempered by critical thinking and empathy, we can ensure that any claims are thoroughly investigated and evaluated. While the discovery of intelligent extraterrestrial life may still be a long way off, the pursuit of this knowledge is a journey worth taking.

Chapter 13 UFOs in Popular Culture

UFOs have been a part of popular culture for decades, inspiring countless books, movies, TV shows, and other forms of media. In this chapter, we will explore the ways that UFOs have been depicted in popular culture and how these depictions have influenced public perceptions of UFOs. Additionally, we will discuss the role that popular culture plays in shaping the public's understanding of UFOs and how investigators can use this to their advantage.

One of the earliest and most enduring depictions of UFOs in popular culture is the 1951 film, The Day the Earth Stood Still. The film tells the story of an alien named Klaatu who comes to Earth to warn humans about the dangers of their own destructive tendencies. Since then, countless films and TV shows have used UFOs as a plot device, ranging from the comedic (Mork & Mindy) to the terrifying (The X-Files).

In addition to films and TV shows, UFOs have also

been the subject of numerous books, both fiction and non-fiction. Some of the most notable works of UFO fiction include The War of the Worlds by H.G. Wells, Close Encounters of the Third Kind by Steven Spielberg, and Communion by Whitley Strieber. Meanwhile, non-fiction works like The Report on Unidentified Flying Objects by Edward Ruppelt and The UFO Experience by J. Allen Hynek have helped to shape the public's understanding of UFOs.

While some depictions of UFOs in popular culture are purely fictional, others are based on real-life sightings and encounters. For example, the 1977 movie Close Encounters of the Third Kind was inspired by real-life accounts of UFO sightings in the late 1960s and early 1970s. Similarly, the TV show The X-Files was heavily influenced by the research of J. Allen Hynek, who served as a technical advisor to the show.

As UFO investigators, it is important to be familiar with the ways that UFOs have been portrayed in popular culture, as these depictions can have a significant impact on the public's perceptions of UFOs. By understanding the cultural context surrounding UFOs, investigators can better communicate their findings to the public and help to dispel common myths and misconceptions.

One of the most persistent myths about UFOs in popular culture is the idea that they are always piloted by extraterrestrial beings. While this is certainly one possibility, there are many other explanations for UFO sightings, including natural phenomena, misidentifications, and hoaxes. By presenting a balanced and nuanced view of UFOs, investigators can help to combat these myths and promote a more evidence-based approach to UFO investigation.

Another important consideration when it comes to UFOs in popular culture is the potential for hoaxes and misinformation. Unfortunately, there are many individuals who seek to exploit the public's fascination with UFOs for personal gain, whether through hoax videos or fraudulent claims of alien contact. Investigators must remain vigilant in their efforts to separate fact from fiction and to maintain a healthy skepticism when evaluating UFO claims.

In addition to its potential pitfalls, popular culture can also be a powerful tool for promoting public interest in UFO investigation. By leveraging the public's fascination with UFOs, investigators can help to generate interest in their work and increase the likelihood of obtaining new sightings and data. However, this must be done with care, as investigators must be careful not to sensationalize their findings or make unfounded claims about extraterrestrial life.

Overall, the study of UFOs in popular culture is an important aspect of UFO investigation. By understanding the ways that UFOs have been depicted in films, books, and other media, investigators can better communicate their findings to the public and help to promote a more evidence-based approach to UFO investigation.

Chapter 14 UFO Investigation Organizations

When it comes to investigating UFOs, it's important to be aware of the various organizations and groups that exist. These organizations have different approaches, beliefs, and methods, and can provide valuable resources for UFO investigators. In this chapter, we'll explore some of the most prominent UFO organizations and what they have to offer.

One of the oldest and most well-known UFO organizations is the Mutual UFO Network (MUFON). MUFON is a nonprofit organization dedicated to the scientific study of UFOs, with a focus on collecting and analyzing UFO sighting reports. MUFON has a large database of UFO sightings and provides training and resources for investigators. They also hold an annual symposium where researchers and enthusiasts gather to discuss the latest developments in the field.

Another notable organization is the Center for the Study of Extraterrestrial Intelligence (CSETI). CSETI is focused on establishing peaceful communication with extraterrestrial civilizations and has developed protocols for initiating contact. They also investigate UFO sightings and have a network of trained investigators.

The National UFO Reporting Center (NUFORC) is a privately funded organization that collects and disseminates information on UFO sightings. They have a large database of sightings and provide a public reporting platform for witnesses to submit their reports. NUFORC also investigates sightings and works with other organizations to analyze data.

The Disclosure Project is a nonprofit organization that aims to disclose information on UFOs and extraterrestrial intelligence. They have collected testimonies from government and military officials who claim to have knowledge of UFOs and cover-ups, and have presented this information to the public and media.

The Scientific Coalition for Ufology (SCU) is a group of scientists, researchers, and investigators who are dedicated to studying UFOs from a scientific perspective. They use a data-driven approach to investigate sightings and develop hypotheses based on evidence.

Other notable organizations include the International UFO Congress, the UFO Research Coalition, and the Aerial Phenomena Investigations Team. Each organization has its own unique approach and focus, but all aim to advance our understanding of UFOs.

When it comes to investigating UFO organizations, it's important to do your research and choose reputable and trustworthy groups. Some organizations may have questionable motives or lack scientific rigor, and it's important to be discerning in your approach.

Once you've identified an organization to work with, it's important to understand their methods and protocols. Some organizations may have specific requirements for submitting sightings or conducting investigations, and it's important to follow these guidelines to ensure your findings are taken seriously.

Working with UFO organizations can provide valuable resources and connections for investigators. By collaborating with other researchers and pooling resources, investigators can increase the likelihood of uncovering new information and advancing our understanding of UFOs.

However, it's important to maintain independence and objectivity in your investigations, even when working with organizations. It's important to remain open-minded and objective, and to follow the evidence where it leads, rather than being swayed by preconceived beliefs or biases.

UFO organizations can be a valuable aspect of UFO investigations. By collaborating with reputable and trustworthy groups, investigators can gain access to valuable resources and connections. However, it's important to maintain independence and objectivity in your investigations, and to be discerning in your approach to working with organizations. By keeping an open mind and following the evidence, we can continue to advance our understanding of UFOs.

Chapter 15 Conducting UFO Investigations

UFO investigations require a great deal of planning and attention to detail. In this chapter, we will cover some of the key steps to conducting a successful investigation.

1. Building a team: Before starting an investigation, it is important to build a team of dedicated and knowledgeable individuals who can work together to gather and analyze evidence. Look for people with diverse skill sets, such as researchers, scientists, and technicians.

2. Developing a plan: Every investigation should start with a solid plan. Determine the scope and objectives of the investigation, and develop a plan for how to collect and analyze evidence.

3. Collecting evidence: There are many types of evidence that can be collected during a UFO investigation, including eyewitness accounts, photographs, videos, and physical traces. It is important to gather as much evidence as possible to support the investigation.

4. Analyzing evidence: Once evidence has been collected, it must be carefully analyzed to determine its validity and relevance to the investigation. This may involve using scientific methods, such as lab analysis, or relying on eyewitness testimony.

5. Interviewing witnesses: One of the most important steps in any UFO investigation is interviewing witnesses. This requires good communication skills and the ability to ask probing questions while remaining objective and neutral.

6. Identifying patterns: By analyzing the evidence and witness accounts, investigators may be able to identify patterns or commonalities that can help to shed light on the nature of the UFO phenomenon.

7. Consulting with experts: UFO investigations often require specialized knowledge and expertise. Consider consulting with experts in fields such as astronomy, physics, and psychology to help interpret and analyze evidence.

8. Maintaining objectivity: It is important for investigators to maintain objectivity and avoid jumping to conclusions. All evidence should be analyzed and interpreted based on its merits, without any preconceived biases.

9. Keeping records: Throughout the investigation, it is important to keep detailed records of all evidence, interviews, and analysis. This will help to ensure that the investigation is transparent and can be

reviewed by others.

10. Collaboration with other investigators: Collaboration with other investigators can be invaluable, as it allows for the sharing of information and perspectives. Consider joining a UFO investigation group or attending conferences to connect with other researchers.

11. Respecting privacy: When conducting an investigation, it is important to respect the privacy of witnesses and refrain from disclosing their personal information without their consent.

12. Evaluating risk: Investigating UFOs can sometimes involve risks, such as encountering dangerous terrain or hostile individuals. Evaluate these risks before conducting an investigation and take appropriate precautions.

13. Conducting field investigations: Field investigations can be especially challenging, as they require investigators to collect evidence in remote or difficult-to-access locations. Plan accordingly and bring appropriate equipment.

14. Documenting procedures: All procedures should be carefully documented to ensure that the investigation is transparent and can be reviewed by others.

15. Reporting findings: Once the investigation is complete, it is important to report the findings in a clear and concise manner. This may involve

publishing a report, presenting at a conference, or submitting a paper for peer review.

16. Staying up-to-date: The field of UFO investigation is constantly evolving, with new evidence and technology becoming available all the time. Stay up-to-date by reading scientific journals, attending conferences, and staying connected with other investigators.

Conducting a successful UFO investigation requires a great deal of planning, collaboration, and attention to detail. By following these key steps and maintaining objectivity throughout the investigation, investigators can help to shed light on this fascinating and mysterious phenomenon.

Chapter 16 Ethics of UFO Investigation

As with any field of study, ethics play a crucial role in UFO investigations. The search for truth must be balanced with the responsibility to respect the privacy and well-being of others. In this chapter, we will explore the ethical considerations that must be taken into account in UFO investigations.

First and foremost, it is important to always obtain consent from individuals before sharing their personal experiences or information publicly. This is especially true when investigating cases of alleged UFO abductions. People who have experienced such events may already feel vulnerable and violated, and it is important to respect their privacy and autonomy.

Another ethical consideration is the use of deception in investigations. While it may be tempting to deceive witnesses or hoaxers to obtain information, doing so violates their trust and may have legal consequences. Instead, investigators should use honest and transparent methods to gather information.

Similarly, investigators should never alter or manipulate evidence to support their theories or conclusions. This not only violates the scientific method, but also undermines the integrity of the investigation and erodes public trust.

In cases where government or military officials may be involved, investigators must also consider the potential repercussions of their actions. Revealing classified information or violating security protocols can have serious consequences, and investigators must weigh the potential benefits of obtaining information against the risks of harm.

It is also important to consider the impact that investigations may have on the wider community. In some cases, sensationalized or inaccurate reporting can lead to fear and panic among the public. Investigators must strive to report information in a responsible and accurate manner, avoiding speculation or conjecture.

Additionally, investigators must take care to avoid perpetuating harmful stereotypes or prejudices. In particular, investigations involving claims of alien abductions or contact should be conducted with sensitivity and cultural competence.

Finally, investigators must always prioritize the safety and well-being of themselves and others involved in the investigation. This may involve taking precautions to ensure physical safety, as well as providing emotional support to witnesses who may be experiencing distress.

Ethical considerations are essential in any UFO investigation. Investigators must balance the desire for truth with the responsibility to respect the privacy, autonomy, and well-being of others. By approaching investigations with integrity and sensitivity, we can promote a more responsible and effective approach to the study of UFOs.

Chapter 17 Communicating UFO Findings

As UFO investigators, it is not enough to simply conduct thorough investigations and analyze data. It is equally important to communicate our findings effectively to various audiences. Whether we are presenting our findings to fellow investigators, the public, or government officials, clear and concise communication is crucial in advancing our understanding of UFO phenomena. In this chapter, we will explore the various ways in which UFO findings can be communicated, the challenges that arise, and the best practices for effective communication.

One of the most common ways of communicating UFO findings is through the publication of reports and articles. These can take the form of official reports from government agencies or independent investigations by private organizations. Regardless of the source, it is important that these reports are written in a clear and concise manner that can be easily understood by a wide range of audiences. Technical jargon and scientific language should be avoided where possible, and any complex concepts should be explained in simple terms.

Another important aspect of communicating UFO findings is through public presentations and lectures. This can take the form of talks given at conferences, public events, or through media interviews. In these cases, it is important to tailor the presentation to the specific audience. Presenting to a group of scientists, for example, would require a different approach than presenting to a group of interested members of the public. In either case, it is important to be clear, concise, and engaging.

In addition to traditional forms of communication, the internet has opened up new opportunities for sharing and discussing UFO findings. Social media platforms, forums, and blogs provide a means for investigators to connect with each other and with the public. However, it is important to be aware of the challenges posed by online communication, such as the spread of misinformation and the difficulty of verifying the accuracy of information.

One of the biggest challenges in communicating UFO findings is the skepticism and even hostility that can be encountered. While there are many individuals and organizations that are open to discussing and investigating UFO phenomena, there are also many who dismiss the topic as nonsense or actively work to discredit investigations. In these cases, it is important to remain calm and rational, and to present evidence and findings in a clear and logical manner.

Another challenge in communicating UFO findings is the sensitive nature of some cases. In cases where there may be national security implications or other sensitive information involved, investigators may be limited in what they can share publicly. In these cases, it is important to respect the boundaries set by government agencies and to maintain the confidentiality of any sensitive information.

Effective communication of UFO findings also requires an understanding of the cultural and societal context in which the investigation is taking place. Beliefs about UFOs vary widely across different cultures, and investigators must be sensitive to these differences in order to effectively communicate their findings. It is also important to consider the potential impact of UFO findings on society, such as the effect on religious beliefs or the potential for widespread panic.

In addition to communicating with the public, effective communication with government officials is also important in advancing our understanding of UFO phenomena. In many cases, government agencies may have access to information or resources that are not available to the public. Establishing positive relationships with government officials can be an important way to gain access to this information and to encourage government agencies to take UFO phenomena seriously.

Finally, it is important to consider the ethical implications of communicating UFO findings. Investigators must always strive to be honest, transparent, and responsible in their communication of findings. This includes being clear about the limitations of the evidence, avoiding sensationalism or exaggeration, and respecting the privacy and confidentiality of individuals involved in investigations. Effective communication of the data is crucial in advancing our understanding of these phenomena. Whether through written reports, public presentations, or online forums, clear and concise communication is essential.

In conclusion, communicating UFO findings is an important aspect of UFO investigations. It is essential to be transparent, honest, and respectful when communicating findings to build trust and credibility with the public. The language used should be clear and concise, and tailored to the audience. By following these guidelines, UFO investigators can effectively communicate their findings and contribute to the understanding of this enigmatic phenomenon.

Chapter 18 The Future of UFO Investigation

As we move forward in our exploration of UFOs, it is important to consider the future of UFO investigation. With the advancements in technology and the increasing public interest in the topic, it is likely that the study of UFOs will continue to evolve and expand in the coming years. In this chapter, we will explore the potential future of UFO investigation and what it may hold for us.

One area that is likely to see significant growth is the use of drones and other advanced technologies in UFO investigation. These tools have the potential to greatly enhance our ability to collect data and analyze sightings. We may also see increased collaboration between different organizations and individuals involved in UFO investigation, leading to more comprehensive and unified approaches to studying the phenomenon.

Another important consideration for the future of UFO investigation is the potential impact of disclosure. If governments or other entities were to release information confirming the existence of extraterrestrial life or advanced technologies, it would likely have a significant impact on the field. It could lead to increased public interest and funding for research, but also potentially create new challenges and ethical considerations.

It is also important to consider the role of public perception in the future of UFO investigation. As the topic becomes more mainstream, there is the potential for increased skepticism and criticism. It will be important for investigators to maintain a high level of professionalism and objectivity in their work in order to maintain credibility.

Another area to consider is the potential for international collaboration and cooperation in UFO investigation. The phenomenon is not limited to any one country, and there is much to be gained by sharing data and insights across borders.

In addition, we may see increased focus on the psychological and sociological aspects of UFO belief and experience. This could involve studying the cultural and societal factors that contribute to belief in UFOs, as well as the potential psychological effects of close encounters or sightings.

The future of UFO investigation may also involve increased focus on exploring other hypotheses beyond the extraterrestrial hypothesis, such as interdimensional or time travel theories. As we continue to expand our understanding of the universe and the nature of reality, it is likely that new possibilities and theories will emerge.

In terms of technology, we may see the development of new tools and methods for collecting and analyzing data, such as advanced sensors or AI algorithms. There may also be increased use of crowd-sourcing and citizen science approaches, allowing for a broader range of individuals to contribute to UFO investigation.

It is important to note that while the future of UFO investigation may hold many exciting possibilities, it is also likely to present new challenges and ethical considerations. Investigators must continue to prioritize accuracy, transparency, and respect for privacy and individual rights in their work.

As we look to the future of UFO investigation, it is important to remain open-minded and willing to adapt as new information and technologies emerge. By maintaining a high level of professionalism and ethical standards, we can ensure that the study of UFOs continues to advance in a responsible and meaningful way.

Chapter 19 UFO Tourism

UFOs have long captured the imagination of people around the world, inspiring countless stories, theories, and investigations. It is no surprise, then, that some people are drawn to places where famous UFO sightings have occurred, or where there are purported to be ongoing UFO activity. This has given rise to a form of tourism known as UFO tourism, where people travel to these locations specifically to seek out evidence of extraterrestrial visitors.

But what exactly is UFO tourism? At its core, it is a type of niche tourism that is focused on exploring places where UFO sightings or encounters have taken place. This can include visiting sites of famous UFO incidents, such as Roswell, New Mexico, or Area 51 in Nevada. It can also involve attending UFO conventions, visiting museums and exhibitions dedicated to UFOs, or even taking part in UFO-themed tours.

The history of UFO tourism can be traced back to the 1940s and 50s, when the first major UFO sightings began to occur in the United States. As more people became interested in UFOs, sightings and reports increased, and with them came an increase in tourism to areas where these sightings had occurred. Today, UFO tourism has expanded beyond the United States to include locations all around the world.

One of the main draws of UFO tourism is the opportunity to visit places where famous or significant UFO sightings have taken place. For example, Roswell, New Mexico, is famous for the alleged crash landing of a UFO in 1947, which has become the subject of numerous books, documentaries, and films. Visitors to Roswell can tour the UFO Museum and Research Center, which provides information about the incident and the ongoing investigation into what happened.

Another popular destination for UFO tourists is Sedona, Arizona, which is known for its purported UFO sightings and vortexes (areas of concentrated energy that are believed by some to have healing or spiritual properties). Visitors to Sedona can take part in UFO tours, which guide them to areas where sightings have occurred, or participate in vortex tours, which take them to the various vortex sites in the area.

While UFO tourism can be a boon to local economies and communities, there are also concerns about its potential impact. Some critics argue that it can exploit local cultures and beliefs, particularly if those beliefs are tied to UFO sightings or encounters. Additionally, safety can be a concern, particularly if tourists venture into remote or dangerous areas in search of evidence.

For those interested in taking part in UFO tourism, there are a number of safety considerations to keep in mind. It is important to research the area ahead of time, and to be aware of any potential dangers or risks. It is also important to respect local cultures and beliefs, and to avoid disrupting the lives of local residents.

UFO tourism can also raise awareness and interest in the field of ufology, leading to more people getting involved in UFO investigations and research. However, it is important to approach UFO tourism with caution and ethical considerations, as it can also attract individuals who may not have the best intentions or who may spread misinformation. Overall, UFO tourism can be a positive aspect of the UFO phenomenon, but it should be approached with a critical eye and a focus on responsible investigation and education.

Chapter 20 The Psychology of UFO Belief

The topic of UFOs has been a fascinating and controversial subject for decades, and with more and more sightings and experiences being reported, the question arises as to why people believe in UFOs. The psychology of UFO belief is a complex and multifaceted issue, influenced by a variety of factors including individual personality traits, cultural and social influences, and personal experiences. In this chapter, we will explore the various psychological factors that contribute to UFO belief and how they impact UFO investigations.

One of the most important factors in the psychology of UFO belief is cognitive biases. Cognitive biases refer to the inherent flaws in human reasoning that can lead to inaccurate or irrational beliefs. For example, confirmation bias is the tendency to seek out information that supports one's pre-existing beliefs while disregarding evidence that contradicts them. In the context of UFOs, this can manifest in individuals who only seek out information that confirms their belief in extraterrestrial visitation while ignoring more plausible explanations for UFO sightings.

Another factor that influences UFO belief is personal experience. Many individuals who believe in UFOs have had their own encounters or sightings, which can lead to a strong conviction in the existence of extraterrestrial life. However, personal experiences are often subjective and open to interpretation, and can be influenced by factors such as expectation, suggestibility, and memory distortion.

Cultural and social influences also play a role in UFO belief. The media and popular culture have portrayed UFOs and extraterrestrial life in a certain way, often sensationalizing and dramatizing the subject. This can lead to a cultural acceptance of the idea that UFOs are evidence of extraterrestrial visitation, and can contribute to the development of a subculture of UFO believers.

Belief in UFOs can also be influenced by individual personality traits. For example, individuals who are more open to new experiences and ideas may be more likely to believe in the possibility of extraterrestrial life. Similarly, those who have a greater need for control and order in their lives may be less likely to accept the idea that UFOs are a mystery that cannot be explained.

The impact of these psychological factors on UFO investigations cannot be underestimated.

Investigators must be aware of their own biases and work to minimize them in order to conduct objective and thorough investigations. Additionally, investigators must also be aware of the impact that cultural and social influences can have on UFO belief, and work to avoid sensationalizing the subject in their investigations.

One important consideration in UFO investigations is the potential for false reporting or hoaxes. Some individuals may report UFO sightings or encounters that are not genuine, either as a prank or for personal gain. Investigators must be able to distinguish between genuine reports and those that are false or misleading, and must approach each case with a healthy dose of skepticism.

Another important consideration in UFO investigations is the potential for psychological distress or trauma in witnesses. Some individuals may experience psychological distress after a UFO sighting or encounter, which can impact their mental health and well-being. Investigators must be sensitive to the potential for psychological distress and provide support and resources for witnesses who may be struggling.

UFO investigations also have ethical considerations, particularly in regards to the privacy and confidentiality of witnesses. Investigators must obtain consent from witnesses before sharing their stories or personal information, and must take steps to protect their privacy and well-being.

Finally, it is important for investigators to approach UFO investigations with a scientific mindset. While belief in the existence of extraterrestrial life is not inherently unscientific, investigations must be conducted using rigorous scientific methods and must be based on evidence rather than speculation or personal beliefs.

UFO investigation is a complex field that requires a diverse range of skills and expertise. From collecting and analyzing data to examining historical and current cases, debunking hoaxes, and exploring different hypotheses, UFO investigators must be equipped with both scientific rigor and an open mind. Additionally, ethics and professionalism are crucial components of conducting investigations that are respectful of witnesses, stakeholders, and the wider community. As we continue to explore the mysteries of UFOs and potential contact with extraterrestrial life, it is important to approach this subject with curiosity, respect, and a commitment to seeking the truth. With the guidance and advice provided in this book, readers will be well-equipped to begin their own investigations and contribute to the ongoing quest for understanding the phenomenon of UFOs.

Chapter 21 The Skeptical Perspective

As a UFO investigator, it's important to approach the subject with an open mind, but also a healthy dose of skepticism. While many people believe in the existence of extraterrestrial life and UFOs, others remain unconvinced. In this chapter, we will explore the skeptical perspective and how it relates to the study of UFOs.

The skeptical perspective is grounded in scientific inquiry and critical thinking. Skeptics believe that extraordinary claims require extraordinary evidence, and that the burden of proof lies with those making the claims. When it comes to UFOs, skeptics argue that there is no concrete evidence to support the notion that they are extraterrestrial in origin.

One of the primary criticisms leveled against the study of UFOs is the lack of scientific rigor in the field. Skeptics argue that UFO researchers often rely on anecdotal evidence and eyewitness testimony, which are notoriously unreliable. They also point out that the scientific method requires the ability to replicate experiments and test hypotheses, something that is difficult to do with UFO sightings.

Another issue raised by skeptics is the prevalence of hoaxes and misidentification. Skeptics argue that many UFO sightings can be easily explained as natural phenomena or man-made objects, and that some sightings are simply hoaxes or publicity stunts. In some cases, skeptics allege that UFO sightings are deliberately staged in order to generate media attention or promote a particular agenda.

Despite these criticisms, there are still many people who believe in the existence of UFOs and extraterrestrial life. Some skeptics argue that the desire to believe in UFOs is rooted in a cultural fascination with science fiction and the unknown. Others suggest that the search for extraterrestrial life is a natural outgrowth of humanity's innate curiosity about the universe.

For those who remain skeptical, there are a number of alternative explanations for UFO sightings. Some suggest that sightings could be attributed to natural phenomena, such as meteorological events or optical illusions. Others point to the possibility of advanced military technology, suggesting that some UFO sightings could be the result of experimental aircraft or surveillance equipment.

Another alternative explanation is the

psychological hypothesis, which suggests that some UFO sightings could be the result of hallucinations or other cognitive biases. This theory suggests that people may see what they want to see, or that their perceptions may be influenced by their beliefs and expectations.

Despite the skepticism surrounding the study of UFOs, there are still many unanswered questions. For example, what causes some people to report UFO sightings while others do not? How can we differentiate between legitimate sightings and hoaxes? And what can we learn from studying UFOs, regardless of their origin?

Perhaps the most important lesson to be learned from the skeptical perspective is the value of critical thinking and scientific inquiry. By approaching the study of UFOs with a healthy dose of skepticism, we can avoid falling prey to hoaxes and misidentification, and ensure that we are collecting and analyzing evidence in a rigorous and systematic way.

At the same time, it's important to remain open-minded and curious about the unknown. While skeptics may argue that there is no concrete evidence to support the existence of UFOs, there are still many unanswered questions about the universe and our place in it. By approaching these questions with an open mind and a willingness to learn, we can continue to expand our understanding of the world around us.

The skeptical perspective provides an important counterbalance to the belief in UFOs and extraterrestrial life. While skeptics may question the validity of many UFO sightings, they also remind us of the importance of critical thinking and scientific inquiry. By approaching the study of UFOs with a balanced perspective, we can continue to learn about the mysteries of the universe, while avoiding the pitfalls of hoaxes and misidentification.

Chapter 22 UFO Hypotheses

The UFO phenomenon has puzzled and intrigued people for decades, and it has inspired a wide range of hypotheses attempting to explain its origin and nature. In this chapter, we will explore some of the most prominent and debated hypotheses about UFOs, from the extraterrestrial to the paranormal.

The Extraterrestrial Hypothesis:
The Extraterrestrial Hypothesis (ETH) suggests that UFOs are spacecraft piloted by intelligent beings from other planets or star systems. This theory gained widespread popularity in the 1950s and 1960s, when UFO sightings increased dramatically, and some of the most famous cases, such as the Roswell incident, were reported. The ETH assumes that these aliens are visiting Earth to study our planet and its inhabitants, and possibly to establish contact or even influence our culture and evolution.

The Interdimensional Hypothesis:
The Interdimensional Hypothesis (IDH) suggests that UFOs are not physical spacecraft but rather vehicles or entities that exist in parallel dimensions

or alternate realities. According to this theory, these entities or vehicles can enter our physical world by manipulating space-time, which explains why they can appear and disappear suddenly and perform seemingly impossible aerial maneuvers. The IDH also suggests that some UFO encounters, such as alien abductions, may involve a transfer of consciousness rather than a physical relocation.

The Time Travel Hypothesis:
The Time Travel Hypothesis (TTH) suggests that UFOs are not from other planets or dimensions but from our own future. According to this theory, time travelers from the future have developed advanced technology that allows them to travel back in time and study past events, including their own history. Some proponents of the TTH suggest that UFO sightings are not sightings of actual spacecraft but rather of time machines, which can explain why UFOs seem to defy the laws of physics and disappear without a trace.

The Secret Military Technology Hypothesis:
The Secret Military Technology Hypothesis (SMT) suggests that some UFO sightings are not caused by extraterrestrial or interdimensional entities but by advanced military aircraft and experimental technology. According to this theory, governments and militaries around the world have developed highly advanced and secret technologies that can explain some of the most puzzling and elusive UFO sightings. The SMT hypothesis suggests that

some governments might be deliberately spreading rumors about extraterrestrial visits to cover up their own technological advancements.

The Paranormal Hypothesis:
The Paranormal Hypothesis suggests that some UFO sightings and encounters are caused by supernatural or paranormal phenomena rather than physical spacecraft or entities. According to this theory, some UFO sightings might be caused by psychic projections or apparitions, while alien abductions might be explained by sleep paralysis, hallucinations, or hypnosis. The Paranormal Hypothesis also suggests that some UFO sightings might be linked to other paranormal phenomena, such as ghosts, poltergeists, or cryptids.

The Cultural and Sociological Significance of UFOs:
Apart from their scientific and technological implications, UFOs also have cultural and sociological significance. The phenomenon has inspired countless works of fiction, art, and popular culture, and it has become a symbol of human curiosity, imagination, and yearning for the unknown. The UFO phenomenon has also been interpreted in different ways by different cultures and societies, reflecting their beliefs, fears, and values. Some societies view UFOs as a sign of impending doom or invasion, while others see them as a source of hope and inspiration.

With all of this in mind, the UFO phenomenon

remains one of the most intriguing and fascinating mysteries of our time. While there are a plethora of theories and hypotheses regarding the origin and nature of UFOs, no single explanation can fully account for all the observed phenomena. It is possible that some UFO sightings may have conventional explanations, while others may remain unexplained. It is important to approach the study of UFOs with an open mind, but also with a healthy dose of skepticism and critical thinking. By continuing to investigate and analyze UFO sightings and experiences, we may eventually uncover the truth behind this enigmatic phenomenon. Regardless of the ultimate explanation, the study of UFOs has significant implications for our understanding of the universe and our place in it, and it serves as a testament to the enduring human curiosity and fascination with the unknown.

Chapter 23 Conclusion

Thank you for reading! This book has provided a comprehensive guide to investigating UFOs. By following the methods outlined in this book, you can collect, analyze, and interpret data to get a better understanding of UFOs. You've learned how to debunk hoaxes, study historical and current cases, and examine various perspectives on UFOs.

It's important to remember that conducting UFO investigations requires discipline, critical thinking, and a willingness to explore new possibilities. The scientific approach, which emphasizes empirical evidence and rational thinking, can be very effective in investigating UFOs. However, it's also important to remain open to spiritual, metaphysical, and cultural perspectives on this phenomenon.

You've explored various hypotheses for UFOs, including the extraterrestrial, interdimensional, time travel, secret military technology, and paranormal hypotheses. Each of these perspectives has its strengths and weaknesses, and it's up to you to evaluate the evidence and decide which one seems most plausible.

In addition to investigating UFOs, we've also examined the sociological and cultural significance of this phenomenon. UFOs have captured the imaginations of people around the world, and studying them can help us better understand our own society and beliefs.

It's also important to remember the ethical considerations involved in UFO investigations. Respect for individuals, privacy, and property should always be maintained. We should also be cautious not to jump to conclusions or make assumptions based on limited evidence.

As we look to the future of UFO investigations, we can expect to see new technologies and methods that will help us better understand this phenomenon. We may also see increased collaboration between government agencies, scientific institutions, and citizen researchers.

Finally, it's important to remember the wonder and excitement that comes with investigating UFOs. The mystery of this phenomenon continues to fascinate and intrigue people around the world. By following the methods outlined in this book and remaining open to new possibilities, you too can contribute to our understanding of this enigmatic and fascinating phenomenon. Happy investigating!

Appendix: Resources for UFO Investigation

In this appendix, we have compiled a list of resources that can be helpful for anyone interested in UFO investigation. These resources include websites, organizations, books, and other materials that can provide additional information and guidance.

1. Websites

- National UFO Reporting Center (NUFORC): A website that collects and shares UFO sighting reports from around the world.
- Mutual UFO Network (MUFON): A nonprofit organization that investigates UFO sightings and provides educational resources.
- Center for UFO Studies (CUFOS): A research organization that studies UFO sightings and related phenomena.
- The Black Vault: A website that provides access to declassified government documents related to UFOs and other topics.

2. Organizations

- Mutual UFO Network (MUFON): A nonprofit organization that investigates UFO sightings and provides educational resources.
- Center for UFO Studies (CUFOS): A research organization that studies UFO sightings and related phenomena.
- The Disclosure Project: An organization that

seeks to disclose information about UFOs and extraterrestrial life.
- The Scientific Coalition for Ufology (SCU): A group of scientists and researchers who study UFO sightings using scientific methods.

3. Books

- "The UFO Experience: A Scientific Inquiry" by J. Allen Hynek: A classic book that provides a scientific perspective on UFO sightings and related phenomena.
- "UFOs and the National Security State" by Richard Dolan: A comprehensive history of government involvement in UFO research and investigation.
- "The Report on Unidentified Flying Objects" by Edward Ruppelt: A firsthand account of the author's experiences investigating UFO sightings for the U.S. Air Force.
- "UFOs: Generals, Pilots, and Government Officials Go on the Record" by Leslie Kean: A collection of firsthand accounts of UFO sightings from government officials and military personnel.

4. Other Materials

- The Condon Report: A 1968 government-funded study on UFO sightings that concluded there was no evidence of extraterrestrial visitation, but acknowledged the need for continued research.
- Project Blue Book: A U.S. Air Force study on UFO sightings that ran from 1947 to 1969 and investigated over 12,000 sightings.

- The SETI Institute: An organization that searches for extraterrestrial intelligence using radio telescopes and other methods.
- The Drake Equation: A formula that estimates the number of intelligent civilizations in the Milky Way galaxy, based on factors such as the number of habitable planets and the likelihood of life emerging on those planets.

UFO Investigation Simple Forms

1. Witness Report Form

Name: ______________________

Contact Information: ______________________

Date: ______________________

Time: ______________________

Location: ______________________

Description of Sighting: ______________________

2. Investigative Notes Form

Date: ______________________

Time: ______________________

Location: ______________________

Description of Investigation: ______________________

Evidence Collected: ______________________

Witness Interviews: ______________________

Additional Notes: ______________________

3. Data Collection Form

Date: ______________________

Time: ______________________

Location: ______________________

Weather Conditions: ______________________

Number of Witnesses: ______________________

Description of Sighting: ______________________

Duration of Sighting: ______________________

Type of Object: ______________________

Color of Object: ______________________

Speed of Object: ______________________

Distance from Observer: ______________________

Direction of Travel: ______________________

4. Photo and Video Documentation Form

Date: ______________________

Time: ______________________

Location: ____________________

Description of Object: ____________________

Type of Camera: ____________________

Camera Settings: ____________________

Number of Photos or Videos Taken: ____________________

Additional Notes: ____________________

These forms can be customized to fit the needs of individual investigators and their investigations. They provide a standardized way of collecting and organizing data, which can be essential for analyzing and interpreting the findings. UFO investigators should also keep in mind the importance of respecting the privacy and confidentiality of witnesses, and obtaining their consent before sharing their information or evidence with others.

In addition to these forms, other useful resources for UFO investigators include databases of UFO sightings and reports, reference books on UFOs and related topics, and online forums or communities where investigators can share their experiences and findings with others. It is also important for investigators to stay up-to-date on the latest developments in the field of UFO research and investigation, and to continue learning and exploring new ideas and hypotheses.

Overall, the appendix serves as a valuable resource for UFO investigators, providing them with the tools and information they need to conduct thorough and professional investigations, and contributing to the ongoing efforts to unravel the mysteries of UFOs and their potential implications for humanity.

Acknowledgements:

This book on UFO investigation and research would not have been possible without the contributions and support of many individuals and organizations. I am grateful to everyone who has shared their knowledge, experiences, and resources with me throughout this project.

First and foremost, I would like to express my deepest gratitude to the UFO Network and its subscribers for their passion and dedication to exploring the mysteries of the universe. Your enthusiasm and curiosity have inspired me and kept me motivated to continue researching and writing about UFOs.

I would also like to extend my sincere appreciation to Heidi-Lore for her valuable insights, feedback, and guidance throughout the writing process. Her knowledge and writing expertise have been invaluable to me and have greatly enhanced the quality of this book.

I would like to thank everyone who has shown an interest in UFOs and has contributed to the advancement of our understanding of these phenomena. Your open-mindedness, curiosity, and willingness to explore new ideas are what make this field so exciting and dynamic.

Additionally, I would like to express my gratitude to the many individuals who have generously shared their UFO experiences and stories with me. Your willingness to share these personal and sometimes profound experiences has given me a deeper appreciation for the complexity and diversity of the UFO phenomenon.

Finally, I would like to acknowledge the many researchers, investigators, and organizations that have contributed to our understanding of UFOs. Their tireless efforts and dedication to this field have paved the way for future generations of researchers and have helped to establish UFO investigation as a legitimate and important area of scientific inquiry.

References

1. Alexander, J. B. (2011). UFOs: myths, conspiracies, and realities. Skyhorse Publishing, Inc.

2. Clark, J. (2012). The UFO book: encyclopedia of the extraterrestrial. Visible Ink Press.

3. Dolan, R. M. (2014). UFOs and the national security state: chronology of a cover-up, 1941-1973. Hampton Roads Publishing.

4. Friedman, S. (2008). Flying saucers and science: a scientist investigates the mysteries of UFOs. New Page Books.

5. Hynek, J. A. (1972). The UFO experience: a scientific inquiry. Ballantine Books.

6. Jacobs, D. M. (1998). The UFO controversy in America. Indiana University Press.

7. Kean, L. (2010). UFOs: generals, pilots, and government officials go on the record. Harmony Books.

8. Keyhoe, D. (1950). The flying saucers are real. Fawcett Publications.

9. Randle, K. D. (1995). The UFO casebook. Warner Books.

10. Ruppelt, E. J. (1956). The report on unidentified flying objects. Doubleday.

11. Salter, F. K. (1997). A history of the military UFO

program. KGRA Publishing.

12. Schuessler, J. A. (2017). The Cash-Landrum UFO incident: a case study. CreateSpace Independent Publishing Platform.

13. Sagan, C. (1996). The demon-haunted world: science as a candle in the dark. Ballantine Books.

14. Saunders, D. R., & Harkins, A. (2015). UFOs? Yes!: Where the Condon Committee Went Wrong. Fund for UFO Research.

15. Stringfield, L. (1977). Situation Red: The UFO Siege!. Fawcett Publications.

16. Trench, B. (1960). The sky people. New American Library.

17. Vallee, J. (1975). The invisible college: what a group of scientists has discovered about UFO influences on the human race. E.P. Dutton.

18. Vallee, J. (1991). Revelations: alien contact and human deception. Ballantine Books.

19. Webb, W. (2016). UFO investigations manual: UFO investigations from 1982 to the present day. Haynes Publishing Group.

20. Zimmermann, R. A. (1980). The Roswell incident. William Morrow and Company.

Made in the USA
Monee, IL
24 June 2023

36955490R00052